AF206956

UNDERSTANDING NATURAL DISASTERS

TORNADOES

by Sue Bradford Edwards

BrightPoint Press

San Diego, CA

© 2025 BrightPoint Press
an imprint of ReferencePoint Press, Inc.
Printed in the United States

For more information, contact:
BrightPoint Press
PO Box 27779
San Diego, CA 92198
www.BrightPointPress.com

LIBRARY OF CONGRESS CATALOGING-IN-PUBLICATION DATA

Names: Edwards, Sue Bradford, author.
Title: Tornadoes / by Sue Bradford Edwards.
Description: San Diego, CA: BrightPoint Press, [2025] | Series: Understanding natural disasters | Includes bibliographical references and index. | Audience: Grades 7-9
Identifiers: LCCN 2024004141 (print) | LCCN 2024004142 (eBook) | ISBN 9781678208721 (hardcover) | ISBN 9781678208738 (eBook)
Subjects: LCSH: Tornadoes--Juvenile literature. | Tornado damage--Juvenile literature.
Classification: LCC QC955.2 .E39 2025 (print) | LCC QC955.2 (eBook) | DDC 363.34/923--dc23/eng/20240216
LC record available at https://lccn.loc.gov/2024004141
LC eBook record available at https://lccn.loc.gov/2024004142

CONTENTS

AT A GLANCE

- A tornado is a rotating column of air. It can last as short as a few minutes or as long as a few hours.

- Tornadoes are spawned by thunderstorms. Not every thunderstorm spawns a tornado.

- It is hard to predict when a tornado will spawn.

- The United States is the country that experiences the most tornadoes. They are most common in an area known as Tornado Alley.

- Small tornadoes do almost no damage. A large tornado can level a building.

- Cleaning up after a tornado can be dangerous. Downed power lines should be avoided until officials declare that the area is safe.

- A tornado watch means the weather is right for severe thunderstorms. A tornado could spawn. People in the area should prepare to take shelter.

- A tornado warning means a tornado has been spotted. People in the area should take shelter immediately.

- People should shelter in a basement room, safe room, or storm shelter. They should stay away from windows.

THE POWER OF TORNADOES

I t was 8:00 p.m. on March 24, 2023, in Rolling Fork, Mississippi. Tracy and Tim Harden were working at their diner, Chuck's Dairy Bar. Tracy had just received a text from her sister. Then she got one from her daughter. Both messages had similar warnings. A tornado had been spotted, and Tracy needed to find shelter right away.

Tracy thought about where to shelter. "Cooler!" she yelled.

If a tornado is visible, then it's time to seek shelter.

The Hardens and seven other people ran to the diner's walk-in cooler. Tim pulled the door open. He shoved people inside. They heard the building breaking apart. Tim stepped inside. The wind yanked the door from his hand. He grabbed the door and

Mobile homes are especially vulnerable to tornado damage.

looked up. The diner's ceiling had been torn away. He closed the door.

The cooler shook. When it stopped shaking, the door was stuck. Tracy called 911. She said they were inside the cooler. When the storm passed, the cooler and a bathroom were all that remained of the diner.

People cleared **debris** from the cooler. They opened the door. Everyone escaped safely.

Tracy Harden was shocked by what she saw outside. Two motel buildings were gone. Trailers in a trailer park had been flattened. Debris lay everywhere. The people in the diner had lived through a tornado.

The National Weather Service has offices across the country where scientists monitor the weather.

SURVIVING A TORNADO

A tornado is a **rotating** cone of fast-moving wind. It moves across the landscape. Some tornadoes are small. They do little damage. They may pull the siding off a building. Other tornadoes are huge. They flatten buildings and flip cars and trucks. They tear trees from the ground.

When a tornado forms, weather warnings save lives. The National Weather Service issued a warning 20 minutes before the tornado reached Rolling Fork. Some people got phone warnings. Tracy Harden got warnings from her family. These warnings helped save the lives of the people in her diner.

CHAPTER ONE

WHAT ARE TORNADOES?

A tornado isn't simply a strong wind. It is a special weather event that has the potential to become a serious natural disaster. Tornadoes don't appear out of nowhere. They form inside storms.

It can be hard to see a tornado's formation. Tornadoes become visible when they contain rain. Tornadoes are also easy to see when they pick up dirt or debris. If a tornado is forming nearby, people need

Weather warning sirens sound a loud alarm when weather gets severe.

to know about it as soon as possible. Taking shelter during a storm can save lives. To be prepared, it helps to know how tornadoes form.

Lightning temporarily heats nearby air to around 50,000°F (28,000°C).

WHAT CAUSES TORNADOES?

The weather pattern that creates tornadoes is called a thunderstorm. Thunderstorms are easy to recognize. These storms produce heavy rain and strong winds. Lightning flashes through the sky. Sometimes it strikes the ground. Lightning heats the air it passes through. Hot air expands. This creates the sound of thunder.

Thunderstorms form when two air masses meet. One is a mass of warm, wet air. This mass is near the ground. The other is a mass of cool, dry air. It flows over the warm air. A current of air begins to flow upward. This is called an updraft. The water in the warm air forms clouds. A thunderstorm is born.

As the air in the updraft climbs, it can change direction. Sometimes it speeds up. This can lead to the formation of a vortex. A vortex is a swirling mass of air inside the storm. One kind of air vortex is called a mesocyclone. It begins as a horizontal tube of rotating air. The updraft of the storm then tilts the mesocyclone vertically. It becomes a rotating column of air that can be up to 6 miles (10 km) across.

A TORNADO FORMS

A mesocyclone extends through the storm clouds from bottom to top. It spins, but it is still not a tornado. Sometimes the base of the mesocyclone becomes tighter and smaller. When this happens, the spinning air speeds up. It extends below the

Mesocyclones form on the back end of thunderstorms.

main cloud. Sometimes this wind sucks in the base of the cloud. The cloud begins to rotate. This is called a funnel cloud. If it touches the ground, it becomes a tornado.

Any thunderstorm can **spawn** a tornado. But most storms never do. The National Weather Service explains, "In the United States, only about five percent of thunderstorms become severe

Sometimes multiple funnel clouds form in the same storm.

and only about one percent produce tornadoes."[1] Most tornadoes are formed by thunderstorms called supercells. A supercell is a thunderstorm that rotates. These storms can be very destructive, even without a tornado. Supercells produce strong winds and hail. They may bring heavy rainfall that can cause flooding. Supercells last longer than other thunderstorms. The tornadoes they spawn may also last longer than other tornadoes.

Thunderstorms and tornadoes depend on air currents. Thunderstorms end when the downdrafts become stronger than the updrafts. A downdraft is when the air current flows down. The warm, wet air no longer rises. No new raindrops form. The thunderstorm slowly ends. Temperature differences between the air masses shrink. The winds slow, and the tornado stops spinning.

LEARNING MORE

Scientists who study the weather are called meteorologists. They are still learning about how tornadoes form. There are several things they look for when they watch storms. One of the first things they look for is a cloud called an inflow band. Inflow bands are long, low **cumulus** clouds. When these clouds are present, it means that the storm is pulling in air. These clouds

Radar towers sometimes have domes that protect equipment inside from the elements.

lie to the south of the storm. If they curve or spiral, they show the storm is rotating. Scientists also look for funnel clouds.

Moving air can be hard to see. Meteorologists don't rely on their eyes alone. They use Doppler **radar**. Doppler radar can measure wind speed. This means the radar can spot the formation of tornadoes.

Tornadoes cause hundreds of millions of dollars in property damage in the United States each year.

Meteorologists often observe supercell thunderstorms. Even these storms rarely spawn tornadoes. Most tornadoes are small. They only last for a few minutes. These tornadoes produce winds of 65 to 100 miles per hour (105–161 kmh). Other tornadoes are much larger. They can last for over an hour. They may produce winds of up to 200 miles per hour (322 kmh). This was the kind of tornado that hit Rolling Fork.

WHERE DO TORNADOES OCCUR?

Tornadoes can occur in many parts of the world. They have been recorded in Europe, Africa, and Asia. They happen in South America, Australia, and New Zealand. Canada reports eighty to one hundred tornadoes a year.

The United States has more tornadoes than any other country. It gets about 1,200 tornadoes a year. Tornadoes can form anywhere in the United States. Tornadoes are most common in an area known as Tornado Alley. Some maps limit this area to the Central Plains. This includes Texas, Oklahoma, Kansas, and Nebraska.

WHICH STATES GOT THE MOST TORNADOES IN 2022?

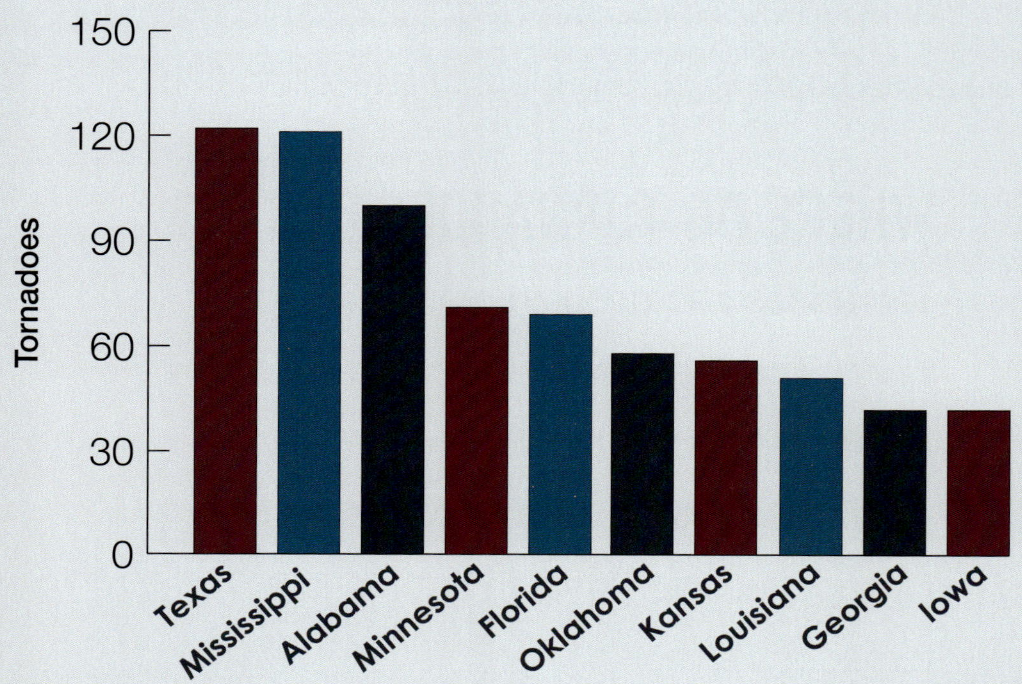

Source: "2022 Annual Final Report Summary," Storm Prediction Center, n.d. www.spc.noaa.gov.

This graph uses US government data to show the ten states that got the most tornadoes in 2022.

Other maps include South Dakota, Iowa, Minnesota, and northern Missouri.

"We see most tornadoes in the central part of the US because the ingredients that are necessary for a tornado come together

there more often than any other place,"
said Harold Brooks.[2] These ingredients are
warm, wet air and colder, dry air. Brooks
is a senior research scientist. He works
at the National Oceanic and Atmospheric
Association (NOAA).

There are other areas where tornadoes
are common, as well. A second danger
zone lies along the Gulf Coast. It includes
Louisiana, Arkansas, Mississippi, Alabama,
Georgia, and Tennessee.

Tornadoes can form at any time of
year. But they are more likely to occur
during certain months. In the early spring,
tornadoes are most common along the
Gulf Coast. From May to June, they are
most common in the Southern Plains
states. These include Texas, Oklahoma,

Tornadoes can form over water. They are called tornadic waterspouts.

and Kansas. June or July is when the Northern Plains and Upper Midwest are most likely to see tornadoes. These regions include North and South Dakota, Nebraska, Iowa, and Minnesota. Tornadoes are most common from 4:00 to 9:00 p.m. But they can form at any time of day. Scientists are still working to learn more about these dangerous storms.

CHAPTER TWO

THE EFFECTS OF TORNADOES

Tornadoes affect people in many ways. The most immediate is the risk of injury during the tornado. Tornadoes can bring down buildings. Heavy items fall to the floor. Glass breaks. Debris flies through the air. People must carefully choose shelters to avoid these dangers. Sturdy objects such as tables can provide cover. People must take special care to protect their heads.

Damaged trees can bring down power lines.

Cleaning up after a tornado can also be dangerous. Tornadoes can tear down power lines. Contact with these lines could cause electrical injuries. The company Duke Energy warns, "Consider all power

lines—as well as trees, limbs and anything else in contact with power lines—energized and dangerous."[3] People should stay away from them until officials have cleaned up the damage.

Buildings damaged by tornadoes could collapse on people. Officials need to examine damaged buildings. They must indicate if the buildings are safe to enter. Then cleanup can begin. During cleanup, people need to dress properly. Sturdy shoes or boots are important. If someone is moving debris, they should wear work gloves. Long sleeves and pants help prevent splinters and cuts. People may need to know how to use power tools. These could include chain saws and trimmers.

Tornadoes can leave survivors with **trauma**. Some people develop post-traumatic stress disorder (PTSD). They may have nightmares. They might avoid people or places that remind them of the tornado. People with PTSD might worry about not being safe. PTSD can also affect one's appetite.

Some people develop depression. Someone who is depressed may not be able to sleep well. They may not want to do things they normally enjoy.

Some people develop anxiety disorders. They become very nervous. Their anxiety could affect their schoolwork. It could affect their relationships with family and friends. It is important for people with PTSD, depression, or anxiety to talk to a doctor.

Volunteers help clean up communities after tornadoes.

COMMUNITY IMPACT

The impact of a tornado goes beyond
individual people. Tornadoes also affect
the communities where they strike. They
destroy roads. They damage homes and
other buildings. They might also damage

the places where people work. In rural areas, tornadoes can flatten fields of crops. In May 2011, a tornado struck Joplin, Missouri. It destroyed more than 8,000 buildings. *USA Today* estimated that the tornado damaged 25 percent of the city.

Communities also suffer from economic damage due to tornadoes. This often happens when businesses are destroyed.

Aid after Tornadoes

The Red Cross and the Federal Emergency Management Agency (FEMA) provide aid after tornadoes. The Red Cross finds shelter for people. It provides food. It provides gloves and shovels for cleanup. The Red Cross also checks homes for damage. FEMA helps pay for housing. It gives people money for household goods such as furniture.

Destroyed businesses might not be able to pay their workers. The businesses might stop selling their products. People must buy them from other businesses. This could cause shortages. When there are shortages, prices rise.

Tornadoes can destroy people's homes. These people must find new places to live. The Rolling Fork tornado struck in March 2023. By July, many people had not yet moved back into their homes. The American Red Cross, a charity, was paying housing costs for 300 people. People were living in hotels in other towns. People who worked in Rolling Fork faced longer driving times. This meant they had to spend more money on gas and car maintenance.

The Red Cross takes blood donations to give to people injured by tornadoes and other disasters.

People spend money where they live. When people must stay far from home, it impacts businesses in their communities. After the tornado, Chuck's Dairy Bar reopened as a food truck. "We need people back home. We need somewhere for them to go," Tracy Harden said. "We are making some money. We have some people working in town. We desperately need to see our hometown people back."[4]

After a tornado, the government
often gives people disaster-relief money.
People might also get money from
their homeowners insurance policies.
Disaster-relief and insurance money is then
spent on recovery. People rebuild their
homes. They buy supplies. Businesses pay
their workers.

ENVIRONMENTAL IMPACT

Tornadoes can harm the environment.
Sometimes tornadoes cause
contamination. This can happen when
they damage pipelines. Pipelines are used
to transport oil and natural gas. These
pipelines stretch across the United States.
A damaged pipeline can leak oil into
the ground.

Tornadoes can also damage storage tanks holding dangerous chemicals. A damaged tank can release these chemicals. The Joplin tornado damaged a valve on a

Oil leaked from pipelines can pollute people's drinking water and harm wildlife.

storage tank. This damaged valve caused a leak of anhydrous ammonia. This chemical is used to help crops grow. It is **toxic** to people. A hazardous materials crew quickly fixed the leak.

Tornadoes may also spread sewage. Leaked chemicals and sewage often pollute water. "This may range from untreated wastewater from a sewage disposal system to chemical contaminants from an industrial facility," said Michael Dean.[5] Dean works at the Oklahoma Environmental Quality Department. Polluted water cannot be used to water crops. People cannot drink it. They cannot bathe in it. The water can be used only after it is treated. People who touch contaminated water need to shower with clean water.

Water treatment plants purify wastewater so that it can be released back into the environment.

Air pollution can also be a serious problem. Tornadoes stir up dust. This dust makes it harder to breathe. Damaged buildings can also cause air pollution. Before the 1980s, a material called asbestos was used in buildings. Particles of asbestos can be released into the air. When breathed, these particles can cause lung diseases.

Tornadoes also destroy trees and other plants. In 2011, a tornado struck near Chico, California. It tore up around 25,000 almond trees. Farmers said it would take 5 years for the trees to regrow. Damage to plants can also lead to soil erosion. This is because the plant's roots no longer hold the soil in place. The soil can blow away. This makes it harder to grow crops.

The rain that comes along with tornadoes can cause flash flooding. Flash floods hit within hours of a storm. They can cause even more damage and pollution.

Thunderstorms also produce lightning. Despite the rain that accompanies these

Flash floods happen in low areas such as underpasses.

storms, lightning strikes can lead to fires.
On August 14, 2023, the Tahoe National
Forest Service reported eight fires caused
by lightning strikes. These fires may be in
remote areas. This means that the fires can
spread before people can reach them.

CHAPTER THREE

DEFENDING AGAINST TORNADOES

People cannot prevent tornadoes. But they can stay safe when one strikes. To do so, they need to know when a tornado is in their area. Weather forecasters help warn people about coming tornadoes. They watch for tornadoes. They use a variety of tools to do this.

Satellites are one tool weather forecasters use. Satellites observe Earth's **atmosphere**. They measure the flow

Some weather satellites are geostationary, which means they hang over the same point on Earth at all times.

of water in the air. This helps weather forecasters see areas where tornadoes might form. They can spot these areas 1 to 7 hours before the storm develops.

Weather forecasters also use Doppler radar. Doppler radar devices point skyward. These devices measure

Weather forecasters use computers to model future weather patterns.

windspeed by tracking the motion of raindrops carried by the wind. Doppler radar can also detect rotation. It is used to observe mesocyclones.

Weather forecasters also look for wind shear. Wind shear is a change in wind with **altitude**. This may occur when wind near the ground moves slower than wind higher up. It could also be a change in the wind's direction. Wind shear can lead to rotation in an updraft.

Predicting a tornado is complicated. Weather forecasters may only be able to give people 10 or 15 minutes of warning. "I would love to be able to tell somebody, 'You know, tomorrow there's going to be a tornado that's going to go through downtown Oklahoma City.' But [. . .] I don't

know if we'll ever be able to get there,"
said Patrick Marsh.[6] Marsh is a weather
forecaster. He works for the NOAA.
He does not think this type of detailed
forecast will ever be possible. He says the
atmosphere is simply too complicated.
There is too much that changes too often.

WEATHER ALERTS

Weather alerts are an important way to
protect people from tornadoes. There
are two types of tornado alerts. These are
tornado watches and tornado warnings.

A tornado watch means people should
be ready for a tornado. The weather
conditions are right in the affected area. A
tornado might form. This is a good time for
people to talk over their plans. They need to

Tornado watches mean that thunderstorms in the area could spawn tornadoes.

know where to shelter. They need to have a way to listen to weather warnings. They also need to have the right supplies. Watch areas are often large. They may include several counties or states.

A tornado warning means people should take shelter immediately. A warning is sounded when a tornado has been seen. Someone may have spotted it in person.

Meteorologists may have seen it on radar. A warning means that there is danger.

People find out about tornado watches and warnings in several ways. Some areas have sirens. The alarms of these sirens can travel over great distances. People who hear them should go inside. They need to

Some areas have community shelters that are open to the public during disasters.

receive a local weather report. They can do this with a radio, television, telephone, or computer. People can also sign up to get text messages. These messages may come from local news stations. Others come from government officials. The messages warn people about the weather. They may also tell people to check local reports.

NOAA weather radios are designed to receive weather warnings. "In rural areas, many people still heavily rely on NOAA weather radios," said Matthew Shpiner.[7] Shpiner works for the University of Miami. He is the director of emergency management. Shpiner reminds people that these radios are a good way to be prepared. He tells people that planning is important.

Emergency kits should be prepared before disaster strikes.

The National Weather Service tells people to keep an emergency kit. The kit should have water and food. The food should be canned so that it will stay fresh. This means people also need a can opener. Kits should include a flashlight and extra batteries. They should have first aid supplies. These include bandages and disinfectant wipes. Antibiotic ointment is important in case of injuries. Other supplies include pet food and blankets. People should have a whistle to signal to rescuers. They need masks and extra clothing. They need toilet paper. They should have rain gear. Emergency kits should be stored in a bag or plastic bin.

TAKING SHELTER

An important part of planning is knowing where to take shelter. Many people shelter in their homes. If people have a basement, this is often best. It should be an area with no windows or outside doors. It should be away from outside walls. A small room or closet can work well. The shelter is where the emergency kit should be stored.

Sometimes people build storm shelters or safe rooms. A storm shelter is outside the house. It can be above ground or underground. It has a heavy metal door. A safe room is built inside the house. Both are made from reinforced concrete that can resist tornado winds. They are strong enough to protect people from debris. Many have ventilation systems. These systems provide fresh air to breathe.

People who are outside during a tornado should take safety measures. The National Weather Service says people should get into a car and drive to a sturdy building. Staying in the car is not safe. Highway overpasses look sturdy. But they can become unsafe if damaged. Sometimes people cannot get to shelter. They should lie flat in a ditch. It is important for them

Basements provide protection from flying debris during tornadoes.

to cover their heads and necks with
their arms.

UNDERSTANDING TORNADOES

A tornado is a dangerous weather event.
It isn't something people can prevent. But
weather forecasters can predict them.
Meteorologists still have a lot to learn
about tornadoes.

Tornadoes don't spawn during most
thunderstorms. Even when tornadoes do
spawn, massive tornadoes are uncommon.

Carried Away

Tornadoes can lift people off the ground. In 2006,
Matt Suter was lifted by a tornado in Missouri. It
carried him 1,307 feet (398 m) from his home. He
survived with minor injuries.

People study weather at colleges and universities before becoming meteorologists.

Most tornadoes are on the ground for only 5 to 10 minutes. Some touch down and do not move. But sometimes a strong tornado causes severe damage to a wide area.

Fortunately, there are steps people can take to be safe. One of the most important steps is to pay attention to tornado watches and warnings. People need to know where to shelter. They should also learn about tornadoes. They might be inspired to join the scientists who study them.

GLOSSARY

altitude
height in relation to ground level

atmosphere
the layer of gases that surround Earth

contamination
the state of being polluted or impure

cumulus
clouds with flat bases and rounded tops

debris
scattered wreckage

radar
technology that measures the speed and position of something using the reflection of waves

rotating
moving in a circle around a center

spawn
to make or produce

trauma
the emotional effects of very disturbing experiences

SOURCE NOTES

CHAPTER ONE: WHAT ARE TORNADOES?

1. "Thunderstorms and Severe Weather Spotting," *National Weather Service*, n.d. www.weather.gov.

2. Quoted in Vox, "Why the US Has So Many Tornadoes," *YouTube*, July 30, 2019. www.youtube.com.

CHAPTER TWO: THE EFFECTS OF TORNADOES

3. "Safety Around Damaged Power Lines," *Duke Energy*, n.d. www.duke-energy.com.

4. Quoted in Scott Simmons, "4 Months after Rolling Fork Tornado, Many Have Yet to Return Home," *WAPT*, July 24, 2023. www.wapt.com.

5. Quoted in Diana Baldwin, "Water Pollution Top Concern after Tornado," *The Oklahoman*, May 5, 1999. www.oklahoman.com.

CHAPTER THREE: DEFENDING AGAINST TORNADOES

6. Quoted in Ari Shapiro, "Scientists Know How Tornadoes Form, but They Are Hard to Predict," *NPR*, May 30, 2019. www.npr.org.

7. Quoted in Robert C. Jones Jr., "The Emotional Impact of Dealing with Deadly Tornadoes," *News @ The U*, March 6, 2019. https://news.miami.edu.

FOR FURTHER RESEARCH

BOOKS

Jack Challoner, *Hurricane & Tornado*. New York: DK, 2021.

Tammy Gagne, *Hurricanes*. San Diego, CA: BrightPoint Press, 2025.

Danielle Haynes, *Tornado Chasers*. New York: PowerKids Press, 2023.

INTERNET SOURCES

"Chasing Tornadoes," *NOVA*, n.d. https://tpt.pbslearningmedia.org.

"How Thunderstorms Form," *UCAR Center for Science Education*, n.d. www.scied.ucar.edu.

Ari Shapiro, "Scientists Know How Tornadoes Form but They Are Hard to Predict," *NPR*, May 30, 2019. www.npr.org.

WEBSITES

American Red Cross
www.redcross.org

The Red Cross helps people affected by a variety of disasters, including tornadoes. Their website has information about how people can help, too.

Kids Discover
https://online.kidsdiscover.com

Kids Discover has a wide variety of information on tornadoes. It includes sections on tornado myths and storm chasers.

NOAA National Severe Storms Laboratory
www.nssl.noaa.gov

The NOAA National Severe Storms Laboratory is run by the US government. Its website gives basic information about tornadoes.

INDEX

IMAGE CREDITS

Cover: © Justin Hobson/Shutterstock Images

5: © Alexey Stiop/Shutterstock Images

7: © Justin Hobson/Shutterstock Images

8: © Simon Annable/Shutterstock Images

10: © Dennis MacDonald/Shutterstock Images

13: © Welland Lau/Shutterstock Images

14: © John D. Sirlin/Shutterstock Images

17: © Laura Hedien/Shutterstock Images

18: © John D. Sirlin/Shutterstock Images

20: © Sergey Sivkov/Shutterstock Images

22: © William A. Morgan/Shutterstock Images

24: © Red Line Editorial

26: © Aramiu/Shutterstock Images

29: © Glynnis Jones/Shutterstock Images

30: © ungvar/Shutterstock Images

33: © Alexey Stiop/Shutterstock Images

36: © Eric Glenn/Shutterstock Images

38: © Kalken/Shutterstock Images

40: © Kletr/Shutterstock Images

42: © Gareth_Bargate/Shutterstock Images

45: © Andrei Armiagov/Shutterstock Images

46: © Dennis MacDonald/Shutterstock Images

49: © Lukas Jonaitis/Shutterstock Images

50: © Sari ONeal/Shutterstock Images

52: © Speedshutter Photography/Shutterstock Images

55: © Jason Finn/Shutterstock Images

57: © Fineart1/Shutterstock Images

ABOUT THE AUTHOR

Sue Bradford Edwards is a nonfiction author who writes in Missouri. She is the author of several BrightPoint Press titles, including *Become a Construction Equipment Operator*, *What Are Learning Disorders?*, and *Robotics in Health Care*. She and her family wait out tornado warnings in a basement room with their weather radio and a variety of camping gear.